KS1 Success

Age 5-7

English

Test

Practice Papers

Rachel Grant

Contents

Introduction and instructions

How these tests will help your child

This book is made up of two complete sets of practice test papers. Each set contains similar test papers to those that your child will take at the end of Year 2 in English reading and English grammar, punctuation and spelling. The tests will assess your child's knowledge, skills and understanding in the areas of study undertaken since they began Year 1. These practice test papers can be used any time throughout the year to provide practice for the Key Stage 1 tests.

The results of both sets of papers will provide a good idea of the strengths and weaknesses of your child.

Administering the tests

- Provide your child with a quiet environment where they can complete each test undisturbed.
- Provide your child with a pen or pencil, ruler and eraser.
- The amount of time given for each test varies, so remind your child at the start of each one how long they have and give them access to a clock or watch.
- You should only read the instructions out to your child, not the actual questions.
- Although handwriting is not assessed, remind your child that their answers should be clear.
- Advise your child that if they are unable to do one of the questions they should to go on to the next one and come back to it later, if they have time. If they finish before the end, they should go back and check their work.

English reading

Paper 1: reading prompt and answer booklet

- Each test is made up of an answer booklet containing two different reading prompts and questions.
- All answers are worth 1 mark, with a total number of 20 marks for each test.
- Your child will have approximately **30 minutes** to read the prompts and answer the questions.
- Read the list of useful words and discuss their meanings with your child.
- Read the practice questions out loud to your child and allow them time to write down their own answer.
- Your child should read and answer the other questions by themselves.

Paper 2: reading answer booklet

- Each test is made up of two different texts and an answer booklet.
- All answers are worth 1 mark (unless otherwise stated), with a total number of 20 marks for each test.
- Your child will have approximately **40 minutes** to read the texts in the reading booklet and answer the questions in the answer booklet.

English grammar, punctuation and spelling

Paper 1: spelling

- Contains 20 spellings, with each spelling worth 1 mark.
- Your child will have approximately **15 minutes** to complete the test paper.
- Using the spelling administration guide on pages 103–104, read each spelling and allow your child time to fill it in on their spelling paper.

Paper 2: questions

- All answers are worth 1 mark (unless otherwise stated), with a total number of 20 marks for each test.
- Your child will have approximately **20 minutes** to complete the test paper.
- Read the practice questions out loud to your child and allow them time to write down their own answer.
- Some questions are multiple choice and may require a tick in the box next to the answer. Some require a word or phrase to be underlined or circled while others have a line or box for the answer. Some questions ask for missing punctuation marks to be inserted.

Marking the practice test papers

The answers and mark scheme have been provided to enable you to check how your child has performed. Fill in the marks that your child achieved for each part of the tests.

Please note: these tests are **only a guide** to the mark your child can achieve and cannot guarantee the same is achieved during the Key Stage 1 tests.

English reading

	Set A	Set B
Paper 1: reading prompt and answer booklet	/20	/20
Paper 2: reading answer booklet	/20	/20
Total	/40	/40

These scores roughly correspond with these standards: up to 10 = well below required standard; 11–20 = below required standard; 21–30 = meets required standard; over 31 = exceeds required standard.

English grammar, punctuation and spelling

	Set A	Set B
Paper 1: spelling	/20	/20
Paper 2: questions	/20	/20
Total	/40	/40

These scores roughly correspond with these levels: up to 10 = well below required level; 11–20 = below required level; 21–30 = meets required level; over 31 = exceeds required level.

When an area of weakness has been identified, it is useful to go over these, and similar types of questions, with your child. Sometimes your child will be familiar with the subject matter but might not understand what the question is asking. This will become apparent when talking to your child.

Shared marking and target setting

Engaging your child in the marking process will help them to develop a greater understanding of the tests and, more importantly, provide them with some ownership of their learning. They will be able to see more clearly how and why certain areas have been identified for them to target for improvement.

Top tips for your child

Don't make silly mistakes. Make sure you emphasise to your child the importance of reading the question. Easy marks can be picked up by just doing as the question asks.

Make answers clearly legible. If your child has made a mistake, encourage them to put a cross through it and write the correct answer clearly next to it. Try to encourage your child to use an eraser as little as possible.

Don't panic! These practice test papers, and indeed the Key Stage 1 tests, are meant to provide a guide to the standard a child has attained. They are not the be-all and end-all, as children are assessed regularly throughout the school year. Explain to your child that there is no need to worry if they cannot do a question – tell them to go on to the next question and come back to the problematic question later if they have time.

Key Stage 1

SET
A

English reading

PAPER 1

English reading

Paper 1: reading prompt and answer booklet

Time:

You have approximately **30 minutes** to complete this test paper.

Maximum mark	Actual mark
20	

First name	
Last name	

Useful words

lighthouse

souvenir

The Story of Grace Darling

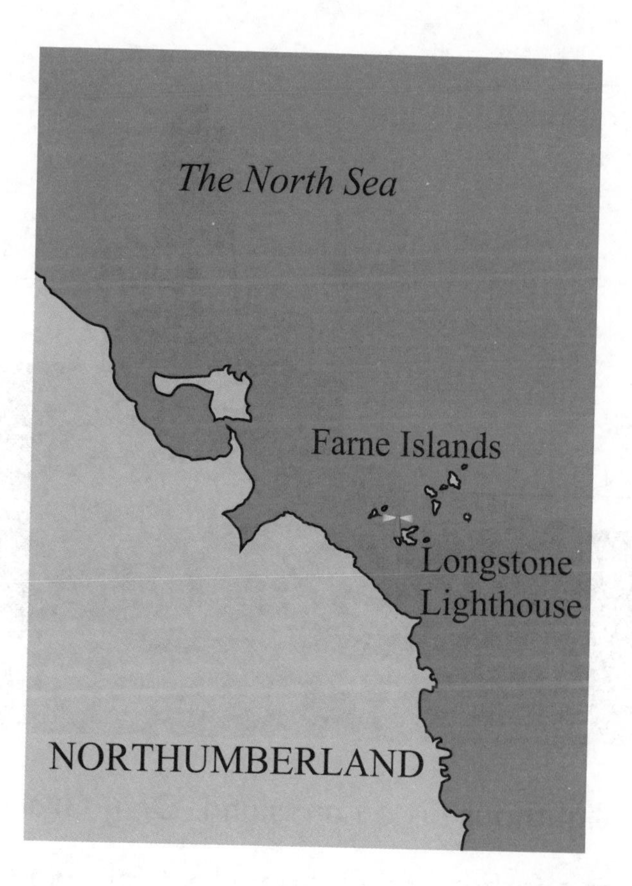

Grace Darling was a lighthouse keeper's daughter. She lived in Longstone Lighthouse, which is in Northumberland in north-east England.

Practice questions

a Who was Grace Darling?

____

b **Find** and **copy** the name of the lighthouse where Grace Darling lived.

____

The lighthouse is on an island. Only Grace's family lived on the island.

Grace's father had an important job. He kept the light burning in the lighthouse.

The light warned ships about dangerous rocks in the sea.

---

1 What did Grace's father do?

____

2 Which word in the text describes what the rocks were like?

Tick **one**.

lonely	☐	important	☐
dangerous	☐	burning	☐

On 7 September 1838 there was a big storm.

A ship was passing Little Farne Island. It had 63 people on board.

Suddenly, the engine stopped. The crew could not use the sails because of the storm.

The ship drifted and hit the rocks.

3 How many people were on board the ship?

Tick **one**.

7 ☐ 63 ☐

18 ☐ 38 ☐

4 Give **two** reasons why the ship drifted and hit the rocks.

1. ______________________________

2. ______________________________

Grace Darling saw the accident. She told her father about it. They did not think anyone on the ship would be alive.

When the sun rose, they saw people clinging to the rocks. Some people were still alive!

---

5 What did Grace do when she saw the accident?

____

6 What did Grace and her father see when the sun rose?

____

Grace and her father took their tiny rowing boat. They rowed through the stormy sea, in winds and rain. They found nine people on the rocks. Grace kept control of the boat while her father went to help.

One by one, they helped the people into the boat. They rowed them to safety.

---

7 **Find** and **copy one** word that tells you what the sea was like.

____

8 Number the pictures 1 to 4 to show the order they happened in the story.

The first one has been done for you.

When people heard Grace Darling's story, she became a national hero. Even Queen Victoria heard the story. Grace was given a silver medal.

Poems and songs were written about Grace. Souvenirs such as postcards and mugs with pictures of Grace on them were made.

---

9 Why did Grace become a national hero?

Tick **one**.

People wrote songs about Grace. ☐

People thought Grace was very brave. ☐

Grace's picture was on postcards. ☐

Queen Victoria heard Grace's story. ☐

Useful words

direction

thousands

Why Owls have Big Eyes

Owl and Pigeon were friends. They liked to meet at dawn and watch the sunrise.

They liked to talk about many things.

One morning, Owl said to Pigeon, "There are more owls than pigeons."

"Nonsense!" said Pigeon. "There are lots more pigeons than owls!"

Practice questions

c *They liked to meet at dawn and watch the sunrise.*

What does *at dawn* mean in this sentence?

Tick **one**.

get together	☐	at home	☐
early in the morning	☐	in the woods	☐

d What did Owl say to Pigeon?

"There are ____"

"There's only one way to find out," hooted Owl. "We will count them!"

"All right," cooed Pigeon. "Where shall we do it? We need a big place."

Owl thought for a minute and then he said, "The Big Wood has lots of trees."

"Fine," said Pigeon. "You tell the owls and I'll tell the pigeons. One week from oday we will count them."

hey both flew off to tell everyone.

---

10 Why was the Big Wood a good place to count the owls and pigeons?

____

11 When did they decide they would count the owls and pigeons?

Tick **one**.

in two weeks	☐	today	☐
later this week	☐	in one week	☐

One week later, the owls arrived at the Big Wood, just as the sun rose.

They flew down from every direction until the trees were full.

They laughed and said, "Too-woo, wah, wah! The pigeons are still asleep!"

They were sure there could not be as many pigeons.

---

12 How do you know there were a lot of owls in the Big Wood?

____

13 *They were sure there could not be as many pigeons.*

What does the word *sure* mean in this sentence?

Tick **one**.

certain ☐ safe ☐

trust ☐ truthful ☐

fter a while they saw huge grey clouds moving towards them. The clouds were igeon wings.

housands of pigeons flew down to the Big Wood and as they landed, the owls oved closer together. Branches broke when too many pigeons tried to land at once.

he owls' eyes grew wider and wider as they tried to see all the pigeons. hey stared and moved their heads from side to side to watch the pigeons.

---

14 **Find** and **copy two** words that tell you what the clouds really were.

____

15 Why did the owls move closer together?

Tick **one**.

To laugh at the pigeons. ☐ To make room for the pigeons. ☐

To watch the pigeons. ☐ To welcome the pigeons. ☐

The pigeons kept coming. The owls could not believe there were so many pigeons!

The owls started to feel nervous and hooted, "Tooooo-woooo! Let's get out of here!"

One by one the owls flew away, up between the branches.

---

16 Why did the owls decide to fly away?

Tick **two**.

It was getting late.	☐	They felt cold.	☐
They felt nervous.	☐	There was no room for them.	☐

17 Number the sentences below from 1 to 4 to show the order they happen in the story.

The first one has been done for you.

The owls flew off.	☐	The owls became worried.	☐
More pigeons arrived.	1	The owls were surprised.	☐

Owl and Pigeon never did count how many owls and pigeons there were.

Now the owls always fly at night so that they will not meet pigeons.

Their eyes are big and round and they stare because they are looking out for pigeons.

---

18 Do you think there were more pigeons than owls?

Tick **one**.

Yes ☐ No ☐

Give one reason for your answer.

____

19 Give **two** things that owls do now because of the pigeons.

1. ______________________________

2. ______________________________

20 Think about all you have read.

How do you think the pigeons felt after the owls flew away, and why?

____

____

END OF TEST

Key Stage 1

SET
A

English reading

PAPER 2

English reading

Paper 2: reading answer booklet

Time:

You have approximately **40 minutes** to read the texts in the reading booklet (pages 82–87) and answer the questions in the answer booklet.

Maximum mark	Actual mark
20	

First name	
Last name	

Questions 1–8 are about
***Why Spiders have Thin Legs* (pages 82–85)**

(page 82)

1 Why did Anansi want to taste the food that his neighbours cooked?

____

(page 82)

2 *they were stubby, but strong.*

What does the word *stubby* mean?

Tick **one**.

black and hairy ☐

short and thick ☐

long and thin ☐

tired and weak ☐

(page 82)

3 Why didn't Anansi wait at Rabbit's house?

____

(page 83)

4 What was Fox cooking for his family?

____

(page 83)

5 At Monkey's house, why did Anansi tie his leg to the oven door?

____

(page 84)

6 **a)** Draw lines to match these characters to the foods they were cooking.

Elephant	nut roast
Squirrel	chicken curry
Giraffe	rice pudding
Hog	honey cake

b) What did each of the animals promise to do when dinner was ready?

____

(page 84)

7 **Find** and **copy** the **three** words that show that Anansi was looking forward to having eight dinners.

____

(pages 82–85)

8 Number the sentences below from 1 to 4 to show the order in which they happened.

The first one has been done for you.

The river washed the webs away.	☐
Rabbit tugged a web.	☐
Anansi talked to Fox.	1
Anansi's legs started to grow long and skinny.	☐

Questions 9–18 are about
***Water* (pages 86–87)**

(page 86)

9 Water is needed for...

Tick **two**.

washing our clothes. ☐

burning wood. ☐

plants to grow. ☐

drying our clothes. ☐

(page 86)

10 Look at the section headed: **The Water Cycle**

Find and **copy** the word that means the same as *small drops*.

____

(page 86)

11 What does the word *cycle* mean in *the water cycle*?

____

____

(page 86)

12 Copy the two labels into the correct boxes on either side of the diagram.

water changes from liquid into gas	water falls as rain, hail, snow or sleet

(page 86)

13 What happens to water when the sun heats it up?

____

____

(page 87)

14 Look at the section: **FACT FILE**

Find and **copy** how much water you have in your:

1. blood ____

2. skin ____

(page 87)

15 Look at the section: **What is Polluted Water?**

Give **three** things that can cause water pollution.

1. ____

2. ____

3. ____

(page 87)

16 **Find** and **copy two** words that mean the same as *rubbish*.

1. ______________________________

2. ______________________________

(page 87)

17 Why is water precious?

Give **one** reason.

____

(page 87)

18 Put ticks in the table to show which sentences are **true** and which are **false**.

One has been done for you.

The information says that...	True	False
humans cause all polluted water.		✓
we should try to save water.		
90% of the world's water is not fresh.		
you should pick up your litter, but only when you are at a beach, lake or river.		
a watering can uses less water than a hosepipe.		

2 marks

END OF TEST

SET A

English grammar, punctuation and spelling

PAPER 1

Key Stage 1

English grammar, punctuation and spelling

Paper 1: spelling

You will need to ask someone to read the instructions and sentences to you. These can be found on page 103.

Time:

You have approximately **15 minutes** to complete this test paper.

Maximum mark	Actual mark
20	

First name	
Last name	

Spelling

Practice question

The children ______________________ in the park.

---

1. The ice is too thin to ______________________ on.

2. Our cat has long ______________________.

3. An elephant's nose is called a ______________________.

4. Sit in the chair ______________________ the door.

5. Sally found a magpie ______________________.

6. Please give me your ______________________ number.

7. We all enjoyed the cricket ______________________.

8. Dad ______________________ Milly's pram.

9. We are going swimming on ______________________.

10. We build a ______________________ on the beach.

11. Hannah has ______________________ to do today.

12. Jack takes the ______________________ to the Post Office.

13. You wear a watch on your ______________________.

14. The hat is ______________________ with yellow flowers.

15. I do not want to ______________________ with you.

16. Please pass me the pink ______________________.

17. This fish is very ______________________.

18. Plastic is a ______________________ material.

19. Our school has a ______________________ in September.

20. Alex gave me an ______________________ to his party.

END OF SPELLING TEST

SET
A

English grammar, punctuation and spelling

PAPER 2

Key Stage 1

English grammar, punctuation and spelling

Paper 2: questions

Time:

You have approximately **20 minutes** to complete this test paper.

Maximum mark	Actual mark
20	....

First name	
Last name	

ractice questions

a Write one word on the line to complete the sentence in the present tense.

Sally is ____________________ a book.

b Tick the correct punctuation mark to complete the sentence.

Where do you live

	Tick **one**.
full stop	☐
question mark	☐
exclamation mark	☐
comma	☐

1 Circle **three** words that should have a capital letter in the sentences below.

it is a hot day. sally and tom are eating ice cream.

2 Tick the correct word to complete the sentence below.

Do you prefer orange juice ________________ apple juice?

and ☐

or ☐

but ☐

so ☐

3 Tick the phrase that uses the correct punctuation to complete the sentence below.

Where are you going ____________________

Tick **one**.

on holiday this year!	☐
on holiday this year.	☐
on holiday this year	☐
on holiday this year?	☐

4 Tick the punctuation mark that should complete each sentence.

Sentence	**Full stop**	**Question mark**
You can have tea at our house		
Are you coming to play football		
Where are my shoes		
It is lunchtime now		

5 What type of word is underlined in the sentence below?

Meera carefully picked up the fluffy hamster.

Tick **one**.

a verb ☐

a noun ☐

an adjective ☐

an adverb ☐

6 Write the missing punctuation mark to complete the sentence below.

What a wonderful surprise

7 Read the sentences below.

Last week, we went to the seaside.

We took our buckets and spades and we packed a picnic.

We played on the sandy beach and in the rock pools.

Then we paddled in the warm sea.

Tick the word that best describes these sentences.

Tick **one**.

questions ☐

commands ☐

exclamations ☐

statements ☐

8 Circle the **four** nouns in the sentence below.

Green plants need air, water and sunlight to grow.

9 Which word correctly completes the sentence?

Brett felt ________________ when he lost his favourite toy.

Tick **one**.

happy ☐

inhappy ☐

unhappy ☐

imhappy ☐

10 Underline the **two** adjectives in the sentence below.

In the garden, a cheeky robin hopped onto a branch and he sang a sweet song.

11 Write **ful** or **less** to complete the word in each sentence.

Did you see that colour_______ rainbow in the sky?

The fear_______ explorer went bravely into the jungle.

12 What type of word is underlined in the sentence below?

The fat ginger cat gazed <u>hungrily</u> at the bright blue fish darting in the water.

an adverb ☐

a noun ☐

an adjective ☐

a verb ☐

13 Write **three** commas in the correct places in the sentence below.

You must bring a coat gloves a hat boots and a packed lunch to school on Friday.

14 The verb in the box is in the present tense.

Write the **past tense** of the verb in the space.

runs

Aisha's mum ____________________ very fast in the egg and spoon race.

15 Why does the underlined word have an apostrophe?

Put your books on the <u>teacher's</u> desk.

____

____

16 Write **was** or **were** to complete the sentence below correctly.

The birds ____________________ flying high up in the sky.

17 **Yusuf and Brandon are at the zoo. They are looking at the parrots. Brandon wants to give the parrots some of his sandwich. Yusuf wants to stop him.**

Write the words that Yusuf says to Brandon in the speech bubble.

Remember to use the correct punctuation.

2 marks

18 The verbs that are underlined are in the present tense.

Write these verbs in the **past tense**.

One has been done for you.

Hari bakes a cake for my birthday.

Hari baked a cake for my birthday.

He uses flour, butter, eggs and sugar.

He ____________________ flour, butter, eggs and sugar.

I choose the chocolate icing and cherries to go on top.

I ____________________ the chocolate icing and cherries to go on top.

2 marks

END OF TEST

SET
B

English reading

PAPER 1

Key Stage 1

English reading

Paper 1: reading prompt and answer booklet

Time:

You have approximately **30 minutes** to complete this test paper.

Maximum mark	Actual mark
20	

First name	
Last name	

Useful words

mainland

million

Two Countries

Greece

Greece is a country in the south-east of Europe.

It has a mainland, which has many mountains, and about 3000 small islands.

Practice questions

a Where is Greece?

____

b **Find** and **copy two** things that you can find in Greece.

1. ____________________

2. ____________________

Wales

Wales is a country in the north-west of Europe. It is part of the United Kingdom.

Wales has 138 mountains and lots of valleys. The longest river in the United Kingdom is called the River Severn. It begins in Wales.

---

1 What is Wales part of?

Tick **one**.

The Severn	☐	The United Kingdom	☐
Europe	☐	The River Severn	☐

2 **Find** and **copy** the name of the country that joins on to Wales.

____

Greece and Wales

These are some differences between Greece and Wales.

	Greece	Wales
Weather in the summer	Usually very hot	Usually cool but sometimes hot
Islands	About 3000	About 50
Highest mountain	Mount Olympus	Mount Snowdon
Capital city	Athens	Cardiff
Population	11 million	3 million
Flag		

3 Tick **two** statements that are **true**.

Greece has more people than Wales. ☐

Wales has more islands than Greece. ☐

Greece's capital city is Athens. ☐

Mount Olympus is the highest mountain in Wales. ☐

4 Write **one** reason why Greece is a popular place to go on holiday.

____

5 Which word in the text means the same as *all the people who live in a country?*

Tick **one**.

people ☐

population ☐

capital ☐

million ☐

Useful words

vanished

electricity

Special Shoes

nce upon a time, not long ago, in a small
llage there lived a little girl called Lena.

ne lived with her mum, her dad and her
der sister Maya. Maya liked running,
ancing, hopscotch and rounders. Lena
ced gardening, drawing and reading
dventure stories.

ractice questions

c Where did Lena live?

____

d Tick to show the things that Lena and Maya liked.

One has been done for you.

	Lena liked	Maya liked
rounders		✓
hopscotch		
reading		
drawing		
dancing		

It was a hot day in July, and Lena and her sister Maya went out to pick raspberries. Soon, Maya decided she was bored with picking. She lay down on the grass and went to sleep.

"Lazybones," said Lena. "I can pick enough raspberries by myself. Mum will be pleased and we can eat them with cream for our tea."

---

6 Why did Lena say "Lazybones"?

Tick **one**.

because Lena was lazy	☐
because Lena was tired	☐
because Lena thought Maya was lazy	☐
because Lena thought Maya was tired	☐

7 Why did Lena want to pick the raspberries by herself? Give **two** reasons.

1. ______________________________

2. ______________________________

All of a sudden, a little old lady appeared from nowhere.

She was carrying a basket.

"Hello Lena," said the old lady. "It's your lucky day!"

"How do you know my name?" asked Lena, surprised. "Why is today lucky?"

"Aha!" said the old lady. "That's for me to know, and for you to find out."

8 Where did the old lady appear from?

____

9 What did the old lady say that surprised Lena the most?

Tick **one**.

"It's your lucky day!" ☐

"Aha!" ☐

"That's for me to know, and for you to find out." ☐

"Hello, Lena." ☐

Lena was curious. "How can I find out?" she asked.

"I see you have some fine raspberries there," said the old lady. "I am very hungry, so here's what I will do. If you give me those raspberries that you have picked, I will give you this shoe box."

"Shoes?" said Lena. "I already have a pair of shoes, thank you very much!"

"Yes," said the old lady, "but not like these. These shoes are special. You'll see!"

---

10 Which word tells you how Lena felt about the old lady?

Tick **one**.

curious	☐	frightened	☐
annoyed	☐	hungry	☐

11 What did the old lady say was inside the shoe box?

____

So Lena gave the old lady the raspberries, and she took the shoe box.

She gasped when she opened the box. Inside were the most beautiful shoes she had ever seen.

They were bright green with bows. The soles were clean and shiny.

"Can I put them on?" she asked the old lady.

But the old lady had vanished.

---

12 How do you know that Lena was surprised when she opened the box?

____ ○

13 How can you tell that the shoes were brand new?

____ ○

14 What happened to the old lady?

____ ○

Lena put the shoes on her feet. To her surprise, they fitted perfectly. As she stood up, she felt a shivery feeling, like icy electricity. It went through her legs and her body, right up to the top of her head. Her hair seemed to stand on end.

How strange! she thought.

"Time to go home!" she called. She poked Maya. "Wakey-wakey, Lazybones!"

---

15 *As she stood up, she felt a shivery feeling, like icy electricity.*

What does the word *shivery* mean in this sentence?

Tick **one**.

warm	☐	very cold	☐
hot	☐	very hot	☐

16 How did Lena feel when she put on the shoes?

Tick **one**.

tired	☐	surprised	☐
bored	☐	upset	☐

17 What did Lena do to wake Maya up?

____

"Where are the raspberries?" said Maya.

"You were a lazybones," said Lena, "so there aren't any."

"You ate them all!" said Maya. "Wait until I tell Mum!"

"Pooh! I bet I get home first!" said Lena, and she whooshed down the hill and zipped through the gate. Maya blinked and her mouth fell open.

Wowee! thought Lena, as she zoomed into the kitchen. These shoes *are* special! I wonder what *else* they can do?

---

18 How do you know that Maya felt amazed when Lena ran home?

____

19 What did Lena wonder when she reached the kitchen?

____

20 Number the sentences below from 1 to 4 to show the order they happened in the story.

The first one has been done for you.

Lena put the shoes on.	
Lena met an old lady.	
Lena's sister fell asleep.	1
Lena traded the raspberries for a shoe box.	

END OF TEST

Key Stage 1

SET
B

English
reading

PAPER 2

English reading

Paper 2: reading answer booklet

Time:

You have approximately **40 minutes** to read the texts in the reading booklet (pages 89–94) and answer the questions in the answer booklet.

Maximum mark	Actual mark
20	....

First name	
Last name	

Questions 1–8 are about
***All about Bees* (pages 89–91)**

(page 89)

1 How many types of honey bee are there?

____

(page 89)

2 Tick **two** things that are **true** about worker bees.

always female ☐

live for up to one year ☐

mate with the queen bee ☐

die after stinging ☐

(page 90)

3 Why do worker bees do the waggle dance?

____

(page 90)

4 Look at the box: **The Bees need you!**

Give **two** reasons why there are fewer bees each year.

1. ____

2. ____

(page 91)

5 Look at the section: **Why do we need Bees?**

Write **two** things that might happen if bees die out.

1. ______________________________

2. ______________________________

(page 90)

6 Look at the section: **Why do we need Bees?**

Find and **copy** the word that means the same as *very important*.

____

(page 91)

7 Look at the section: **Why do we need Bees?**

Number the steps 1 to 4 to show how pollination works.

The first one has been done for you.

Pollen helps other flowers to make seeds.	
The bee moves to other flowers.	
The pollen rubs off the bee's body.	
Pollen from a flower sticks to the bee's body.	1

(page 91)

8 Look at the section: **Make a Bee Garden.**

Draw lines to match the words below to their meaning.

Words	**Meanings**
protect	making flowers
habitats	natural homes
flowering	keep safe

Questions 9–18 are about
***Alien On Board* (pages 92–94)**

(page 92)

9 Who was Mrs Rose?

____

(page 92)

10 Why was life in space lonely?

____

(page 92)

11 How did Skylar, Zoe and Mrs Rose know something had happened?

____

(page 92)

12 *It was the strangest little creature I had ever seen.*

What does the word *strangest* mean?

Tick **one**.

the most furry ☐

the most odd ☐

the most strong ☐

thr most friendly ☐

(page 92)

13 How did Zoe feel about the alien?

____

(page 93)

14 **Find** and **copy** **one** word that tells you how the alien was feeling.

____

(page 93)

15 How do you know Skylar was not frightened of the alien? Give **two** things.

1. ______________________________

2. ______________________________

(page 93)

16 Why couldn't they keep the alien?

____

(page 94)

17 **a)** Put ticks in the table to show which sentences are **true** and which ones are **false**.

The first one has been done for you.

The story says that Mr and Mrs Smik…	**True**	**False**
have five long sticks.		✓
have tentacles.		
have very little fur.		
have pale pink skin.		
have lumpy skin.		

2 marks

b) What sort of creatures do you think Mr and Mrs Smik were?

____

(page 94)

18 Why did Skylar think dog-dog would be back some day?

____

END OF TEST

Key Stage 1

SET
B

English grammar, punctuation and spelling

PAPER 1

English grammar, punctuation and spelling

Paper 1: spelling

You will need to ask someone to read the instructions and sentences to you. These can be found on pages 103–104.

Time:

You have approximately **15 minutes** to complete this test paper.

Maximum mark	Actual mark
20	

First name	
Last name	

Spelling

Practice question

The birds ______________________ into the sky.

---

1. Gina shows great ______________________ on the piano.

2. The postman brought a big ______________________.

3. The eggs ______________________ under the hen.

4. Our new ______________________ loves to play with us.

5. The children had an amazing time at the ______________________.

6. The sky was ______________________ when we left home.

7. Tom ______________________ TV after he has done his homework.

8. Dad is ______________________ vegetables for dinner.

9. Gran is coming to tea on ______________________.

10. Jane was too ______________________ to join in our game.

1. He thinks reading is ____________________ than writing.

2. Eating fresh fruit every day is good for your ____________________.

3. The hot air balloon rose ____________________ the rooftops.

4. She loves all vegetables, especially ____________________.

5. Mum ____________________ to answer the door.

6. Jane ____________________ at school today.

7. Uncle Toby took his ____________________ to the park.

8. The river runs through a deep ____________________.

9. Yesterday I saw a ____________________ with seven spots.

10. I enjoy reading stories with lots of ____________________.

END OF SPELLING TEST

SET
B

English grammar, punctuation and spelling

PAPER 2

Key Stage 1

English grammar, punctuation and spelling

Paper 2: questions

Time:

You have approximately **20 minutes** to complete this test paper.

Maximum mark	Actual mark
20	

First name	
Last name	

Practice questions

a Tick the word that completes the sentence.

Gerbils are usually ____________________ than hamsters.

smallest ☐

small ☐

little ☐

smaller ☐

b Write the words **do not** as one word, using an apostrophe.

Please ____________________ pick the flowers.

1 Underline the noun phrase in the sentence.

Do you like my new green hat?

2 Circle **three** words that must have capital letters in the sentence.

My birthday is in april and jane's birthday is in may.

3 Tick the best option to complete the sentence.

Will you put the books away, please

full stop ☐

exclamation mark ☐

question mark ☐

comma ☐

4 Write the missing punctuation mark to complete the sentence below.

Our school trip is to Leeds Castle

5 What type of word is hurt in the sentence below?

Shelley slipped on the ice and hurt her leg.

	Tick **one**.
a noun	☐
an adjective	☐
an adverb	☐
a verb	☐

6 Which sentence has the correct punctuation?

Tick **one**.

What a great idea ☐

What a great idea! ☐

What a great idea? ☐

What a great idea. ☐

7 Circle the adverb in the sentence below.

We all cheered loudly when my sister won the race.

8 Tick the word that completes the sentence.

They ____________________ at a lovely hotel.

Tick **one**.

stayed ☐

stays ☐

staying ☐

to stay ☐

9 Tick to show whether each sentence is in the **present tense** or the **past tense.**

	Present tense	**Past tense**
I was talking to Mary.		
Jack is brushing his teeth.		

10 Read the sentences below.

Oh no! The dog has chewed my trainers!

Tick the word that best describes the sentences.

statements ☐

exclamations ☐

commands ☐

questions ☐

11 Tick the sentence that is a **statement**.

Tick **one**.

How exciting! We're going on holiday! ☐

Have you got the tickets? ☐

Carry your suitcase to the car. ☐

I have locked the front door. ☐

12 Write **two** commas where they should go in the sentence below.

Spring summer autumn and winter are the four seasons of the year.

13 Tick the sentence that is written correctly in the past tense.

Tick **one**.

We are going to the shop and buying some ice cream. ☐

We went to the shop and bought some ice cream. ☐

We go to the shop and bought some ice cream. ☐

We went to the shop and buy some ice cream. ☐

14 Circle **two** adjectives in the sentence below.

When we woke up the next morning, the soft, white snow was falling silently from the sky.

15 Write the words in the boxes as one word, using an apostrophe.

One has been done for you.

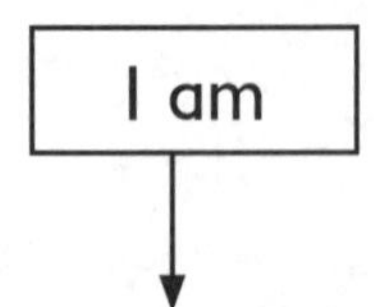

I'm building a sandcastle.

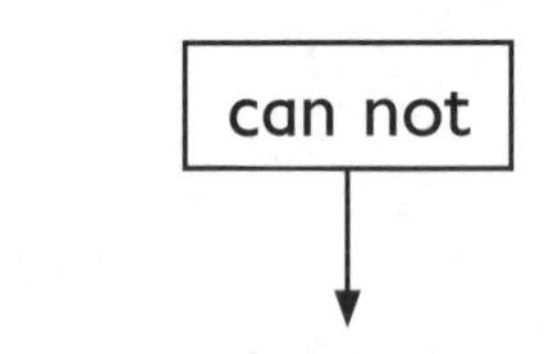

He ______________ go swimming today.

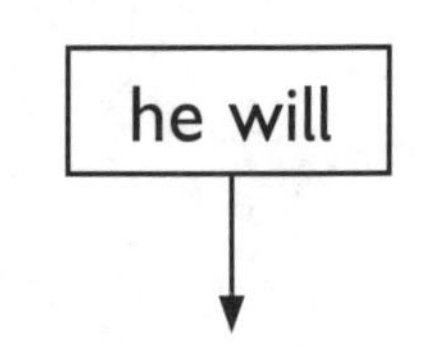

I hope ______________ be better tomorrow.

16 Tick **two** sentences that are correct.

They is closing the paddling pool. ☐

The park is closing soon. ☐

Peter is closing the door quietly. ☐

I are closing my eyes and counting to ten. ☐

17 Draw lines to match the sentences with their correct type.

When we got home it was time for bed.	question
Pass the salt.	exclamation
When is it lunchtime?	statement
Thank you! I love it!	command

18 The verbs in the boxes are in the **present tense.**

Write the verbs in the correct tense in the sentences below.

One has been done for you.

bark

The children were shouting and the dog __was barking__ **at the door.**

try

Mum ______________ **to work but she couldn't think because of the noise.**

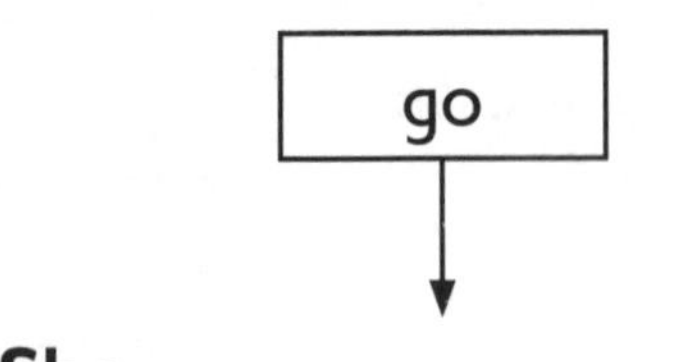

She ______________ **downstairs to find out what was wrong.**

2 marks

END OF TEST

Paper 2:
reading booklet

Why Spiders have Thin Legs

pages 82–85

Water

pages 86–87

Why Spiders have Thin Legs

Once upon a time, a long time ago, there lived a greedy spider named Anansi. Anansi loved his wife's cooking as she always cooked delicious food. Anansi also loved tasting the food that his neighbours cooked for themselves.

One day, Anansi was walking down the road on his short, thick legs. He had eight legs, and they were stubby, but strong. "I wonder what we are having for dinner," Anansi thought. "I hope it is beef stew! Yum yum. Beef stew is my favourite!"

Just then he smelled something delicious. The smell came from Rabbit's house. Rabbit was his good friend. "What's cooking?" asked Anansi. "I smell carrot soup!"

Rabbit came out of his house. He knew Anansi was greedy.

"You are welcome to some soup," said Rabbit, "but it is not ready. If you wait for a while, you can eat with me."

"Oh, no," said Anansi. "I'm sorry, Rabbit, but I can't wait because I am very busy." In fact, Anansi was lazy. He knew that Rabbit would give him jobs to do. Anansi didn't want to sweep the floor or wash dishes. He wanted to eat the carrot soup!

"I have an idea," said Anansi. "I will spin a web. I will tie one end around my leg and one end to your pot. When the soup is ready, you give the web a tug, and I will come back!"

"Good idea!" said Rabbit.

So Anansi spun a web. He tied one end to one of his legs. He tied the other end to the pot.

"See you later!" he said to Rabbit.

Moments later, Anansi smelled another delicious smell. This time it was beans and rice. His friend Fox was cooking a delicious meal for his family. Fox invited Anansi to stay but Anansi was afraid of the jobs he would need to do. He didn't want to do any work, so he tied another web to one of his legs. He tied the other end to the cooking pot.

"Please tug when the beans and rice are ready!" he said.

A few minutes later, Anansi's nose started to twitch. He smelled something delicious!

The smell came from Monkey's house.

"What is that delicious smell?" Anansi called.

"Banana bread!" said Monkey. "It's almost ready. Do you want to taste it?"

"I would love to," said Anansi.

And again, Anansi spun a web. He tied one end to his leg and the other end to the oven door. Monkey thought this was a great idea. He promised to tug on the web when the banana bread was ready.

Then Anansi went on his way.

At Elephant's house, he smelled sweet and creamy rice pudding. Rhino was cooking chops and greens. Squirrel was making a delicious nut roast. Giraffe was busy cooking a wonderful honey cake. Hog was making chicken curry for his family. Each one invited Anansi to stay to dinner. Each one promised to tug on a web when dinner was ready.

At last, Anansi was walking along with a web tied to each of his eight short legs. He was very pleased with himself. He was going to have eight dinners, and do no work at all! His mouth watered as he thought about the delicious meals waiting for him.

Just then, he felt a tug on one of his legs. "It's Rabbit!" cried Anansi.

"Great! Carrot soup!" He turned to go to Rabbit's house. Then he felt a tug on a different leg. Monkey's banana bread was ready!

"Oh dear!" said Anansi. Which dinner should he eat first? Then there was a tug on another leg, and then another. Soon, every one of Anansi's legs was being pulled at once.

"Help!" he cried. His legs were growing thinner and thinner!

What could he do? Just then he had an idea. He jumped as hard as he could, and landed in the river.

The river washed the webs away and his legs were free.

When he climbed out of the river, Anansi was very tired.

He looked down at his legs. They were thin and very long.

"Oh, my!" he said. "What will my wife say?"

He hurried home to dinner.

"What happened?" asked his wife. "Your legs are so long and skinny!"

"Never mind," said Anansi. "I am very hungry. I need some beef stew!"

"You are too late," said his wife. "It's all gone."

So Anansi didn't eat any dinner that day. That is why he still visits everyone's house, searching for food, with his eight thin legs.

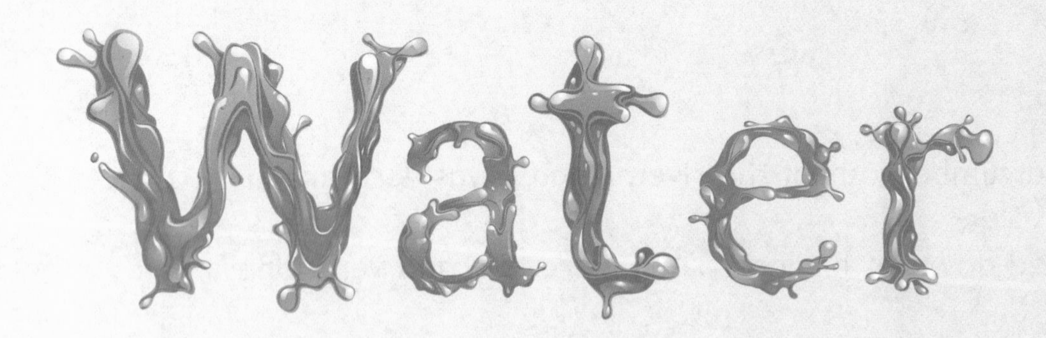

We use water every day. We drink water. We use it to wash our bodies, our clothes and our dishes.

Plants need water to grow and animals need it to live.

The Water Cycle

Water goes round and round, in a cycle. Cycle means a circle of events that happen again and again in the same order.

When the sun heats up water in seas and rivers, some of the water droplets move into the air.

The water changes from a liquid into a gas. Just like the water that comes out of a kettle when it boils!

The water droplets join up to make a cloud.

When the clouds get big and full of water, the water droplets fall from the sky. They may fall as rain or hail, snow or sleet, if it is very cold.

The water falls onto the land, sea and rivers. Or it falls on your umbrella, your roof or your head!

REMEMBER

The water cycle is always happening, all the time. Water is always turning from a liquid into a gas and back again. Clouds are always forming. There is always rain, hail, snow or sleet falling somewhere in the world.

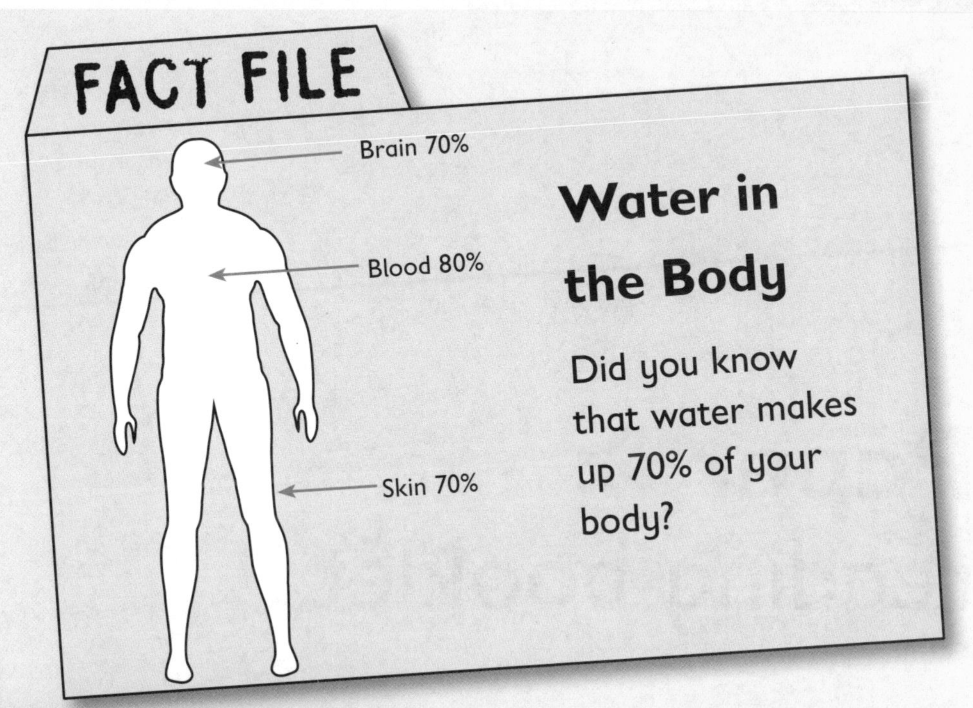

What is Polluted Water?

Polluted water is dirty water that is harmful to many living things.

Sometimes water becomes polluted because of things like storms, floods and animal waste. Some polluted water is caused by human activity. Food waste, chemicals such as oil, and refuse can cause water pollution.

Polluted water can make humans and other animals sick.

Only 1% of the world's water is fresh and safe to drink. The rest is salty and we can't drink it. Water is precious because we need it every day, so use it wisely.

What can you do to help?

- Always pick up your litter, especially when you are at a beach, lake or river.
- Save water by taking shorter showers, turning off taps and using a cup when you brush your teeth.
- Use a watering can (not a hosepipe) to water the garden.

Paper 2:
reading booklet

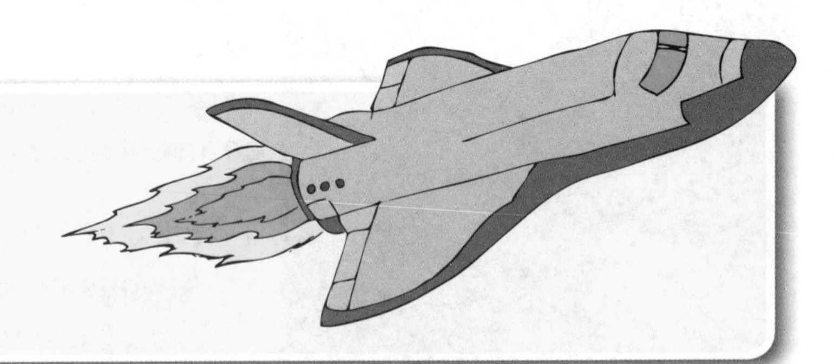

All about Bees

Honey bees live in hives. There are three types of honey bees in the hive: the queen bee, the drones and the workers.

The **queen bee** lives between 1 and 4 years. She eats royal jelly that the worker bees provide. Her job is to lay eggs.

The **drones** are male bees. They mate with the queen to create more bees. They do not have a sting.

The **workers** are female bees. There may be as many as 60000 in one hive. Worker bees collect pollen and nectar, feed the drones, build and protect the hive, collect water and feed the eggs. If they use their sting, they die.

The Waggle Dance

If a worker bee finds flowers with a lot of nectar and pollen, she goes back to the hive. She moves in a figure of eight and waggles her body to show the other bees where the food is. This movement is called the waggle dance.

The Bees need you!

Bees have been disappearing from our country. Every year there are fewer bees. Some of the reasons for this are use of chemicals, loss of habitat, lack of food and poisons in the air.

Why do we need Bees?

Bees do essential work. They take pollen from plant to plant. This is called pollination. The plants need the pollen in order to grow new seeds. Without pollination, our flower, fruit, nut and vegetable plants would die.

Pollination

The flower is colourful and smells good!

It has nectar and pollen in it! Yum!

Yum yum! I have pollen on my body as well as in my mouth!

Oh yummy! The pollen from my body is helping this flower to make seeds!

Make a Bee Garden

We need to protect our bees by making wild habitats for them to live in.

Attract bees to your garden by planting flowers that bees like. They like wild flowers such as cornflowers, poppies, lavender, foxgloves and bluebells.

Fruit trees, herbs and flowering shrubs are also good.

Alien on Board

Our family lives on a spaceship. My dad is the captain of the ship and he's a Galaxy Map Maker. He makes maps of the galaxies we travel through. I'm Skylar and I'm 7 years old. My little sister Zoe and I have lived here all our lives. We live here alone – except for our teacher, Mrs Rose. I almost forgot Mrs Rose because she isn't really a person. She is a computer, but she is very lifelike, and very strict, just like all teachers.

Life in space can be very lonely. It's so quiet, and so dark, except for the stars. Sometimes I wish I had a friend, or that we could meet an alien, for company. But nothing ever happens. We've been travelling in space for years but we have never seen any aliens, monsters or scary creatures at all.

Then, last Tuesday, just as we were finishing our lessons with Mrs Rose, we heard a buzz and a loud CRASH! It came from the landing area. We rushed to see what had happened. There was a small, grey space shuttle, with an alien inside!

The alien looked at us. We looked at the alien. It was the strangest little creature I had ever seen. It was almost completely black, except for its nose, which was brown. It had brown stripes down both of its front legs and it had two eyes! Surely one eye is enough for anyone? And fur should be pink, like ours, not black!

"What on earth!?" said Dad.

"I don't like it!" said Zoe. "Make it go away!"

"Don't be silly," said Dad. "It's more frightened of you than you are of it."

He opened the shuttle door.

The alien crouched down. It looked terrified. Then it showed its teeth, huge and white.

I went up to the alien and knelt down.

"Don't be afraid," I said, gently. "We won't hurt you!"

"Be careful, Skylar!" said Dad.

"It's all right," I replied. "See. It's friendly."

The alien was shaking its bottom. A long, soft, furry stick-like thing made swooshing noises. A soft, pink, wet thing came out of its mouth and brushed against my ear.

"Aw!" I said. "It's cute! Can we keep it, Dad?"

Dad was looking at the shuttle. "Property of Mr and Mrs Smik.
If found, please return to 92, The Hutchings, Planet Zogg," he read.

Zoe was petting the alien, "Cutie, cutie!"

"What sort of alien is it?" I asked Dad

"It's not an alien," said Dad. "It's a dog. I wonder how it got here.
We need to take it home."

"Dog-dog," said Zoe. "Nice dog-dog!"

"Oh!" I said. I was disappointed. Here was my first alien, and it wasn't an alien at all!

"You'd better take it to the main deck," said Dad. "I'll radio to Planet Zogg and speak to Mr and Mrs Smik."

Dog-dog followed us to the main deck.

"Do you think it likes moss cake?" I asked Zoe.

It was fun having dog-dog on the spaceship. First it ate some moss cake. Then it tried the fungus pie. After that it bounced on Zoe's space hopper. Finally, it jumped right onto the control panel and started pressing buttons.

"Stop that!" shouted Dad. "I'm flying this spaceship, not you! We're flying to Planet Zogg, right now!"

When we landed on Planet Zogg, dog-dog's owners were there to meet us. They looked every bit as strange as dog-dog. For a start, they had hardly any fur at all. Their skin was smooth and a pale pink colour all over. They had no tentacles, just four long sticks with five wriggly things at each end. Mr Smik had a patch of red hair just below his nose. Mrs Smik's mouth seemed to have something wrong with it. It was huge and red, like an exploded cherry.

"Darling!" said Mrs Smik. Her voice was squeaky and high.

Dog-dog jumped up at her.

"Oh, you naughty boy!" said Mrs Smik. "Did you fly our shuttle, all by yourself?"

Dog-dog said nothing, but as he turned to go, I'm sure he winked at me. I am almost certain he will be back some day, to eat more fungus pie.

Answers and mark scheme

Set A English reading

Paper 1: reading prompt and answer booklet

The Story of Grace Darling

a a lighthouse keeper's daughter

b Longstone Lighthouse

1 He kept the light burning in the lighthouse. **(1 mark)**

2 lonely ☐ important ☐
dangerous ☑ burning ☐ **(1 mark)**

3 7 ☐ 63 ☑
18 ☐ 38 ☐ **(1 mark)**

4 Any two from: the engine stopped; the crew could not use the sails; because of the storm.
(1 mark: two correct for 1 mark)

5 Grace/She told her father about it. **(1 mark)**

6 Any answers that refer to people clinging to the rocks/people still alive. **(1 mark)**

7 stormy **(1 mark)**

8

Picture	Picture
2	1
3	4

(1 mark: award 1 mark for the pictures correctly numbered)

9 People wrote songs about Grace. ☐
People thought Grace was very brave. ☑
Grace's picture was on postcards. ☐
Queen Victoria heard Grace's story. ☐
(1 mark)

Why Owls have Big Eyes

c get together ☐
at home ☐
early in the morning ☑
in the woods ☐

d more owls than pigeons

10 Any one from: (because) the Big Wood/it had lots of trees; (because) they needed a big place. **(1 mark)**

11 in two weeks ☐ today ☐
later this week ☐ in one week ☑ **(1 mark)**

12 Because the trees were full. **(1 mark)**

13 certain ☑ safe ☐
trust ☐ truthful ☐ **(1 mark)**

14 pigeon wings **(1 mark)**

15 To laugh at the pigeons. ☐
To make room for the pigeons. ☑
To watch the pigeons. ☐
To welcome the pigeons. ☐ **(1 mark)**

16 It was getting late. ☐ They felt cold. ☐
They felt nervous. ☑ There was no room for them. ☑
(1 mark: award 1 mark for two correct answers ticked)

17 The owls flew off. 4 The owls became worried. 3
More pigeons arrived. 1 The owls were surprised. 2
(1 mark: award 1 mark for correct order)

18 Any of the following answers are acceptable.
Yes, because:
- the owls were surprised to see so many pigeons.
- the owls felt nervous because there were so many pigeons.
- the owls flew away because there were more pigeons than owls.
- there were so many pigeons that they were like a huge grey cloud. **(1 mark)**

19 They/Owls fly at night; They/Owls stare; They/Owls look out for pigeons.
(1 mark: two correct for 1 mark)

20 Any plausible, text-based answers are acceptable, e.g. They felt (very) happy/delighted/overjoyed, because:
- the owls had flown off/left.
- they had won the contest.
- there were more pigeons than owls.
- they had shown/proved there are more pigeons than owls. **(1 mark)**

Set A English reading

Paper 2: reading booklet

Why Spiders have Thin Legs

1 Any justified reason, derived from the text, e.g.

- Anansi was greedy.
- Anansi loved food/delicious food. **(1 mark)**

2 black and hairy [ ]
short and thick [✓]
long and thin [ ]
tired and weak [ ] **(1 mark)**

3 Any one of the following:

- he was very busy/lazy
- he didn't want to do any jobs/work/sweep the floor/wash the dishes. **(1 mark)**

4 beans and rice (Accept a delicious meal) **(1 mark)**

5 So Monkey could tug when the banana bread was ready. **(1 mark)**

6 a)

Animal	Food
Elephant	rice pudding
Squirrel	nut roast
Giraffe	honey cake
Hog	chicken curry

(1 mark)

b) To tug on the web when the banana bread was ready. **(1 mark)**

7 His mouth watered. (Accept Anansi's mouth watered.) **(1 mark)**

8 The river washed the webs away. [4]
Rabbit tugged a web. [2]
Anansi talked to Fox. [1]
Anansi's legs started to grow long and skinny. [3]

(1 mark: award 1 mark for correct order)

Water

9 washing our clothes. [✓]
burning wood. [ ]
plants to grow. [✓]
drying our clothes. [ ]

(1 mark: award 1 mark for two correct answers ticked)

10 droplets **(1 mark)**

11 a circle of events that happen again and again in the same order. **(1 mark)**

12 water falls as rain, hail, snow or sleet

water changes from liquid into gas

(1 mark: award 1 mark for both correct labels)

13 Any one of the following: water droplets move into the air; water changes from a liquid into a gas. **(1 mark)**

14 blood 80%; skin 70%
(1 mark: both correct for 1 mark)

15 Any three of the following: storms; floods; animal waste; food waste; chemicals such as oil; refuse
(1 mark: three correct answers for 1 mark)

16 Any two of the following: waste; refuse; litter
(1 mark: two correct answers for 1 mark)

17 Any one of the following: only 1% of the world's water is fresh and safe to drink; most of the world's water is salty and we can't drink it; we need water every day. **(1 mark)**

18

The information says that...	True	False
humans cause all polluted water.		✓
we should try to save water.	✓	
90% of the world's water is not fresh.		✓
you should pick up your litter, but only when you are at a beach, lake or river.		✓
a watering can uses less water than a hosepipe.	✓	

(2 marks for four correct answers, 1 mark for two or three correct answers)

Set A English grammar, punctuation and spelling

Paper 1: spelling

These are the correct spellings:

1. skate
2. claws
3. trunk
4. nearest
5. feather
6. telephone
7. match
8. pushes
9. Wednesday
10. sandcastle

11 nothing
12 parcels
13 wrist
14 purple
15 quarrel
16 chalk
17 bony
18 useful
19 festival
20 invitation

(20 marks: 1 mark for each correct spelling)

Set A English grammar, punctuation and spelling

Paper 2: questions

a Any suitable present tense verb ending in -ing, e.g. reading, holding, carrying, buying.

b full stop ☐
question mark ☑
exclamation mark ☐
comma ☐

1 (it) is a hot day.(sally)and(tom)are eating ice cream.
(1 mark: all three correctly circled for 1 mark)

2 and ☐ or ☑
but ☐ so ☐ **(1 mark)**

3 on holiday this year! ☐
on holiday this year. ☐
on holiday this year ☐
on holiday this year? ☑ **(1 mark)**

4

Sentence	Full stop	Question mark
You can have tea at our house	✓	
Are you coming to play football		✓
Where are my shoes		✓
It is lunchtime now	✓	

(1 mark: all correct for 1 mark)

5 a verb ☐ a noun ☑
an adjective ☐ an adverb ☐ **(1 mark)**

6 What a wonderful surprise! **(1 mark)**

7 questions ☐ commands ☐
exclamations ☐ statements ☑ **(1 mark)**

8 Green (plants)need(air),(water)and(sunlight)to grow.
(1 mark: all four correct for 1 mark)

9 happy ☐ inhappy ☐
unhappy ☑ imhappy ☐ **(1 mark)**

10 In the garden, a cheeky robin hopped onto a branch and he sang a sweet song.
(1 mark: both adjectives underlined for 1 mark)

11 Did you see that colour**ful** rainbow in the sky?
The fear**less** explorer went bravely into the jungle.
(1 mark: two correct suffixes added for 1 mark)

12 an adverb ☑ a noun ☐
an adjective ☐ a verb ☐ **(1 mark)**

13 You must bring a coat, gloves, a hat, boots and a packed lunch to school on Friday.
(1 mark: three correctly placed commas for 1 mark)

14 ran (Accept was running) **(1 mark)**

15 Any response that explains that the apostrophe shows belonging or possession, e.g.

- it shows belonging/possession
- it shows that a/the desk belongs to a/the teacher
- because a/the desk belongs to a/the teacher.

Responses that meet the criteria but use different phrasing should also be marked as correct.

Do not accept responses referring to other uses of the apostrophe, e.g.

- because a letter/letters have been missed out
- because it is short for is/it is. **(1 mark)**

16 were **(1 mark)**

17 Award 2 marks for any command sentence that starts with a capital letter and ends with an appropriate punctuation mark, e.g.

- No, don't do that, Brandon!
- That's a bad idea!
- Stop it, Brandon.

Award 1 mark for any command sentence that does not start with a capital letter and/or that does not end with a full stop or an exclamation mark, e.g.

- no don't do that brandon
- you must not do that

Responses that meet the criteria but use different phrasing should be marked as correct.

Correct spelling is not required for the award of the mark. Although correct sentence punctuation is required for the award of both marks, pupils are not required to use capital letters correctly for Brandon or correct internal punctuation. **(2 marks)**

18 used (Accept was using, had used or has used)
chose (Accept had chosen or have chosen)
(2 marks: 1 mark for each correct answer)

Set B English reading

Paper 1: reading prompt and answer booklet

a in the south-east of Europe

b Any two from: a mainland; mountains; small islands.

1 The Severn ☐
The United Kingdom ☑
Europe ☐
The River Severn ☐ **(1 mark)**

2 England **(1 mark)**

3 Greece has more people than Wales. ☑
Wales has more islands than Greece. ☐
Greece's capital city is Athens. ☑
Mount Olympus is the highest mountain in Wales. ☐
(1 mark: award 1 mark for two correct answers ticked)

4 Any reference to usually hot weather in summer, e.g.
- the weather is usually very hot in the summer
- because it's usually hot in summer. **(1 mark)**

5 people ☐ population ☑
capital ☐ million ☐ **(1 mark)**

Special Shoes

c in a small village

d

	Lena liked	Maya liked
rounders		✓
hopscotch		✓
reading	✓	
drawing	✓	
dancing		✓

6 because Lena felt lazy ☐
because Lena felt tired ☐
because Lena thought Maya was lazy ☑
because Lena thought Maya was tired ☐
(1 mark)

7 Any **two** of the following:
- (because) she could pick enough (by herself)
- (because her) Mum would be pleased
- (so that) they could all eat them with cream for tea.

(1 mark)

8 nowhere **(1 mark)**

9 "It's your lucky day!" ☐
"Aha!" ☐
"That's for me to know, and for you to find out." ☐
"Hello Lena." ☑
(1 mark)

10 curious ☑ frightened ☐
annoyed ☐ hungry ☐ **(1 mark)**

11 special shoes **(1 mark)**

12 She gasped. **(1 mark)**

13 Any answer that quotes or paraphrases the following:
- the soles were clean and shiny, e.g.
  - o the soles were new
  - o the soles were not worn. **(1 mark)**

14 She/the old lady vanished. (Accept she/the old lady disappeared) **(1 mark)**

15 warm ☐ very cold ☑
hot ☐ very hot ☐ **(1 mark)**

16 tired ☐ surprised ☑
bored ☐ upset ☐ **(1 mark)**

17 Any one of the following:
- she poked Maya/her
- she called, "Time to go home!"
- she told her it was time to go home
- she said, "Wakey-wakey, Lazybones!"
- she called her Lazybones and told her to wake up.

(1 mark)

18 Any answer that quotes or paraphrases the following:
- because Maya/she blinked
- because Maya's/her mouth fell open.

(1 mark)

19 Any one of the following points:
- what else they/the shoes could do
- what/other things they/the shoes could do **(1 mark)**

20 Lena put the shoes on. 4
Lena met an old lady. 2
Lena's sister fell asleep. 1
Lena traded the raspberries for a shoe box. 3
(1 mark: award 1 mark for correct order)

Set B English reading

Paper 2: reading booklet

All about Bees

1 three (Accept 3) **(1 mark)**

2 always female ✓
live for up to one year ☐
mate with the queen bee ☐
die after stinging ✓

(1 mark: award 1 mark for two correct answers ticked)

3 To show the other bees where the food is. (Accept to show other bees where there are flowers with a lot of nectar and pollen) **(1 mark)**

4 Any two of the following: use of chemicals; loss of habitat; lack of food; poisons in the air **(1 mark)**

5 Any two of the following: plants would not be pollinated; some plants would not grow new seeds; some/flower/fruit/nut/vegetable plants would die **(1 mark)**

6 essential **(1 mark)**

7 Pollen helps other flowers to make seeds. 4
The bee moves to other flowers. 2
The pollen rubs off the bee's body. 3
Pollen from a flower sticks to the bee's body. 1

(1 mark: award 1 mark for correct order)

8

Words	Meanings
protect	keep safe
habitats	natural homes
flowering	making flowers

(1 mark)

Alien on Board

9 Any one of the following: their/a teacher; a computer; not really a person (Do not accept very lifelike, strict) **(1 mark)**

10 Any one of the following: it was very quiet; it was very dark; it was very dark, except for the stars; they had never seen any aliens/monsters/creatures. **(1 mark)**

11 They heard a buzz and a loud crash. (Accept they heard a loud noise/loud noises) **(1 mark)**

12 the most furry ☐
the most odd ✓
the most strong ☐
the most friendly ☐ **(1 mark)**

13 Zoe/she did not like it. (Accept Zoe/she was frightened of it. Do not accept Zoe/she wanted it to go away.) **(1 mark)**

14 terrified **(1 mark)**

15 Any two of the following: she went up to it; she knelt down; she spoke to it gently; she told it not to be afraid/she said, "Don't be afraid"; she said "We won't hurt you!" **(1 mark: two correct answers for 1 mark)**

16 Any reference to the alien belonging to Mr and Mrs Smik, e.g.
- it/the alien belonged to Mr and Mrs Smik
- it/the alien was the property of Mr and Mrs Smik
- its/the alien's owners were Mr and Mrs Smik.

Also accept:
- they need to take it/the alien home to its owners
- it/the alien does not belong to Skylar. **(1 mark)**

17 a)

The story says that Mr and Mrs Smik...	True	False
have five long sticks.		✓
have tentacles.		✓
have very little fur.	✓	
have pale pink skin.	✓	
have lumpy skin.		✓

(2 marks: two or three correct for 1 mark, four correct for 2 marks)

b) Any reference to human beings, e.g.
- they were human beings
- they were humans
- they looked like humans
- they were probably human beings. **(1 mark)**

18 Any reference to dog-dog winking at Skylar, e.g.
- because she is sure dog-dog winked at her
- because she thinks dog-dog winked at her
- because she thinks dog-dog will be back to eat more fungus pie. **(1 mark)**

Set B English grammar, punctuation and spelling

Paper 1: spelling

These are the correct spellings:

1 skill
2 parcel
3 hatch
4 puppy
5 circus
6 clear
7 watches

8 chopping
9 Thursday
10 shy
11 easier
12 health
13 above
14 cabbage
15 hurries
16 wasn't
17 nephew
18 valley
19 ladybird
20 action

(20 marks: 1 mark for each correct spelling)

Set B English grammar, punctuation and spelling

Paper 2: questions

a smallest ☐ small ☐
most small ☐ smaller ☑

b don't

1 Do you like my new green hat?
(1 mark: all words correctly underlined for 1 mark)

2 My birthday is in april and jane's birthday is in may.
(1 mark: three capital letters identified for 1 mark)

3 full stop ☐
exclamation mark ☐
question mark ☑
comma ☐ **(1 mark)**

4 Our school trip is to Leeds Castle. **(1 mark)**

5 a noun ☐ an adjective ☐
an adverb ☐ a verb ☑ **(1 mark)**

6 What a great idea ☐
What a great idea! ☑
What a great idea? ☐
What a great idea. ☐ **(1 mark)**

7 We all cheered loudly when my sister won the race.
(1 mark)

8 stayed ☑ stays ☐
staying ☐ to stay ☐ **(1 mark)**

9

	Present tense	Past tense
I was talking to Mary.		✓
Jack is brushing his teeth.	✓	

(1 mark: both ticked correctly for 1 mark)

10 statements ☐ exclamations ☑
commands ☐ questions ☐ **(1 mark)**

11 How exciting! We're going on holiday! ☐
Have you got the tickets? ☐
Carry your suitcase to the car. ☐
I have locked the front door. ☑
(1 mark)

12 Spring, summer, autumn and winter are the four seasons of the year.
(1 mark: both commas correctly placed for 1 mark)

13 We are going to the shop and buying some ice cream. ☐
We went to the shop and bought some ice cream. ☑
We go to the shop and bought some ice cream. ☐
We went to the shop and buy some ice cream. ☐
(1 mark)

14 When we woke up the next morning, the soft white snow was falling silently from the sky.
(1 mark: both adjectives circled for 1 mark)

15 He can't go swimming today.
I hope he'll be better tomorrow.
(2 marks: 1 mark for each correct sentence)

16 They is closing the paddling pool. ☐
The park is closing soon. ☑
Peter is closing the door quietly. ☑
I are closing my eyes and counting to ten. ☐
(1 mark: two ticked correctly for 1 mark)

17
When we got home it was time for bed. — statement
Pass the salt. — command
When is it lunchtime? — question
Thank you! I love it! — exclamation

(1 mark)

18 was trying (Accept tried; had tried)
went (Accept was going; had gone)
(2 marks: 1 mark for each correct answer)

Spelling test administration

Ensure that your child has a pen or pencil and a rubber to complete the paper.
Your child is not allowed to use a dictionary or electronic spell checker.
Ask your child to look at the practice spelling question. Do this question together.
For each question, read out the word that your child will need to spell correctly.
Then read the whole sentence.
Then read the word again.
Your child needs to write the word into the blank space in the sentence.
Here is the practice question:
The word is **play**.
The children **play** in the park.
The word is **play**.
Check that your child understands that 'play' should be written in the first blank space.
Explain that you are going to read 20 sentences. Each sentence has a word missing, just like the practice question.
Read questions 1 to 20 to your child, starting with the question number, reading out the word followed by the sentence, and then the word again.
Leave enough time (at least 12 seconds) between questions for your child to attempt the spelling. Do not rush, as the test time of 15 minutes is approximate and children will be given more time if they need it.
Tell your child that they may cross out the word and write it again if they think they have made a mistake.
You may repeat the target word if needed.

Set A English grammar, punctuation and spelling

Paper 1: spelling

Practice question: The word is **play**.
The children **play** in the park.
The word is **play**.

Spelling 1: The word is **skate**.
The ice is too thin to **skate** on.
The word is **skate**.

Spelling 2: The word is **claws**.
Our cat has long **claws**.
The word is **claws**.

Spelling 3: The word is **trunk**.
An elephant's nose is called a **trunk**.
The word is **trunk**.

Spelling 4: The word is **nearest**.
Sit in the chair **nearest** the door.
The word is **nearest**.

Spelling 5: The word is **feather**.
Sally found a magpie **feather**.
The word is **feather**.

Spelling 6: The word is **telephone**.
Please give me your **telephone** number.
The word is **telephone**.

Spelling 7: The word is **match**.
We all enjoyed the cricket **match**.
The word is **match**.

Spelling 8: The word is **pushes**.
Dad **pushes** Milly's pram.
The word is **pushes**.

Spelling 9: The word is **Wednesday**.
We are going swimming on **Wednesday**.
The word is **Wednesday**.

Spelling 10: The word is **sandcastle**.
We build a **sandcastle** on the beach.
The word is **sandcastle**.

Spelling 11: The word is **nothing**.
Hannah has **nothing** to do today.
The word is **nothing**.

Spelling 12: The word is **parcels**.
Jack takes the **parcels** to the Post Office.
The word is **parcels**.

Spelling 13: The word is **wrist**.
You wear a watch on your **wrist**.
The word is **wrist**.

Spelling 14: The word is **purple**.
The hat is **purple** with yellow flowers.
The word is **purple**.

Spelling 15: The word is **quarrel**.
I do not want to **quarrel** with you.
The word is **quarrel**.

Spelling 16: The word is **chalk**.
Please pass me the pink **chalk**.
The word is **chalk**.

Spelling 17: The word is **bony**.
This fish is very **bony**.
The word is **bony**.

Spelling 18: The word is **useful**.
Plastic is a **useful** material.
The word is **useful**.

Spelling 19: The word is **festival**.
Our school has a **festival** in September.
The word is **festival**.

Spelling 20: The word is **invitation**.
Alex gave me an **invitation** to his party.
The word is **invitation**.

Set B English grammar, punctuation and spelling

Paper 1: spelling

Practice question: The word is **flew**.
The birds **flew** into the sky.
The word is **flew**.

Spelling 1: The word is **skill**.
Gina shows great **skill** on the piano.
The word is **skill**.

Spelling 2: The word is **parcel**.
The postman brought a big **parcel**.
The word is **parcel**.

Spelling 3: The word is **hatch**.
The eggs **hatch** under the hen.
The word is **hatch**.

Spelling 4: The word is **puppy**.
Our new **puppy** loves to play with us.
The word is **puppy**.

Spelling 5: The word is **circus**.
The children had an amazing time at the **circus**.
The word is **circus**.

Spelling 6: The word is **clear**.
The sky was **clear** when we left home.
The word is **clear**.

Spelling 7: The word is **watches**.
Tom **watches** TV after he has done his homework.
The word is **watches**.

Spelling 8: The word is **chopping**.
Dad is **chopping** vegetables for dinner.
The word is **chopping**.

Spelling 9: The word is **Thursday**.
Gran is coming to tea on **Thursday**.
The word is **Thursday**.

Spelling 10: The word is **shy**.
Jane was too **shy** to join in our game.
The word is **shy**.

Spelling 11: The word is **easier**.
He thinks reading is **easier** than writing.
The word is **easier**.

Spelling 12: The word is **health**.
Eating fresh fruit every day is good for your **health**.
The word is **health**.

Spelling 13: The word is **above**.
The hot air balloon rose **above** the rooftops.
The word is **above**.

Spelling 14: The word is **cabbage**.
She loves all vegetables, especially **cabbage**.
The word is **cabbage**.

Spelling 15: The word is **hurries**.
Mum **hurries** to answer the door.
The word is **hurries**.

Spelling 16: The word is **wasn't**.
Jane **wasn't** at school today.
The word is **wasn't**.

Spelling 17: The word is **nephew**.
Uncle Toby took his **nephew** to the park.
The word is **nephew**.

Spelling 18: The word is **valley**.
The river runs through a deep **valley**.
The word is **valley**.

Spelling 19: The word is **ladybird**.
Yesterday I saw a **ladybird** with seven spots.
The word is **ladybird**.

Spelling 20: The word is **action**.
I enjoy reading stories with lots of **action**.
The word is **action**.